This book
belongs to

....

ZIGZAG FACTFINDERS

INVENTIONS

Written by
Chris Oxlade

Edited by
Philippa Moyle

zigzag

The consultant, Dr John Becklake, has worked for many years at the Science Museum in London. He has consulted on a large number of books for children.

ZIGZAG PUBLISHING

Published by Zigzag Publishing,
a division of Quadrillion Publishing Ltd.,
Godalming Business Centre, Woolsack Way,
Godalming, Surrey GU7 1XW, England.

Series concept: Tony Potter
Managing Editor: Nicola Wright
Production: Zoë Fawcett
Designed by Iain Ashman and Kate Buxton
Illustrated by Peter Bull, Kuo Kang Chen, Peter Dennis,
Mainline Design, Jeremy Gower and Andy Miles/
Maggie Mundy Agency
Cover illustration: John Fox

Color separations: RCS Graphics, Leeds and Pixel Tech, Singapore
Printed in Singapore

Distributed in the U.S. by SMITHMARK PUBLISHERS
a division of U.S. Media Holdings, Inc.,
16 East 32nd Street, New York, NY 10016

Copyright © 1997 Zigzag Publishing. First published 1994.

ISBN 0-7651-9320-5
8045

Contents

Sources of power 4

Moving ahead 6

Sea and sailing 8

Flying machines 10

Looking closer 12

Weaving a fabric 14

Telling the time 16

Writing and printing 18

Record and playback 20

Keeping in touch 22

Photography and television 24

Electronics 26

Revolution in the home 28

Weird and wonderful 30

Index 32

About this book

Have you ever wondered what life would be like without cars, computers or light bulbs? All of these everyday things are the result of inventors' hard work, and many only appeared in the past one hundred years.

However, not all inventions are new! Did you know that fax machines were first thought of in 1843 or that Egyptians used water to tell the time?

From the wheel to the silicon chip, this book is packed with fascinating facts about amazing, and sometimes even life-saving, inventions that have been made throughout history.

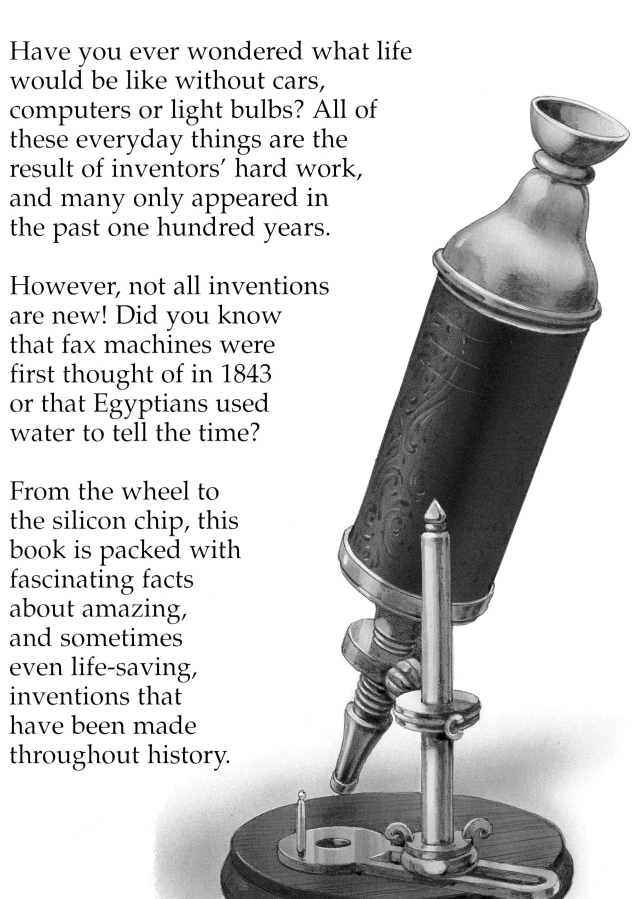

Over 2,000 years ago the Ancient Greeks were using **water wheels** for grinding flour. Water was the main source of power for industry until steam engines were invented.

The weight of water falling into the buckets turned the wheel.

Before engines were invented nature was the only source of power available. Animals pulled carts, and the wind and running water moved windmills and water wheels. Water, wind and animal power are still important today.

However, engines are an important source of power. Cars, trucks, trains, aircraft and ships all have their own special engines to power them along.

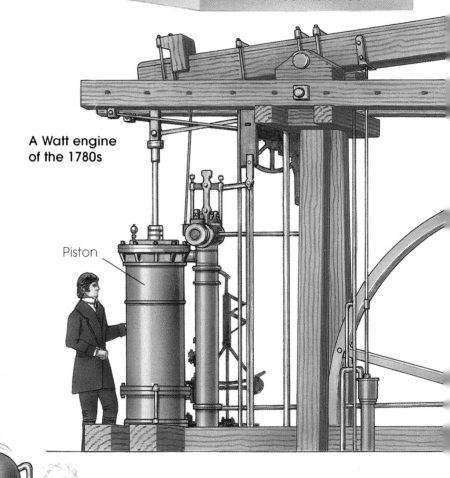

A Watt engine of the 1780s

Piston

The first **steam machine** was made before A.D. 100 by a Greek engineer, called Hero. It spun around as steam shot out of the pipes.

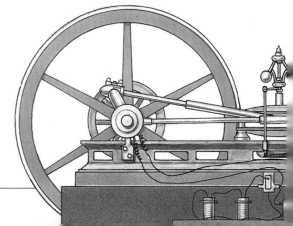

A **turbine** spins very fast when water flows through it. The turbine was invented in 1827. It soon replaced the water wheel.

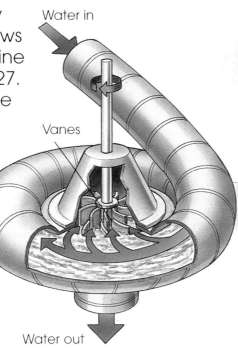

Water in

Vanes

Water out

Water pours into the turbine through a narrow pipe. It pushes the vanes round.

Windmills were first used around 650. They turned huge millstones which ground grain to make flour. They also pumped water and worked machinery.

In 1776 James Watt built a **steam engine** for pumping water out of coal and tin mines. Steam from boiling water moved a piston in and out. The moving piston worked the water pump.

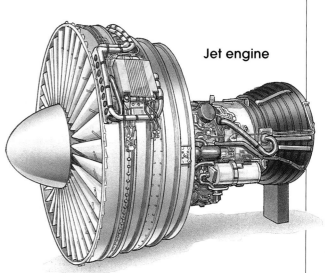

Jet engine

The sort of engine used in most modern cars is called an **internal combustion engine**. The first of these engines was built in 1860.

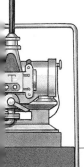

An internal combustion engine has cylinders and pistons like a steam engine. The first one used gas for fuel.

In 1939 the first aircraft with a **jet engine** took off. It was called the Heinkel He 178. Jet engines meant that aircraft could fly much faster than before.

The wheel was one of the most important inventions in history. Think how difficult it would be to get around without it. Cars, bicycles, trains and carts use wheels.

For thousands of years carriages and carts were pulled by horses. But as soon as engines were invented people began making powered vehicles.

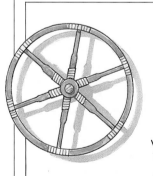

Wheels were invented over 5,000 years ago. They were made from solid wood. About 4,000 years ago wheels with spokes were invented.

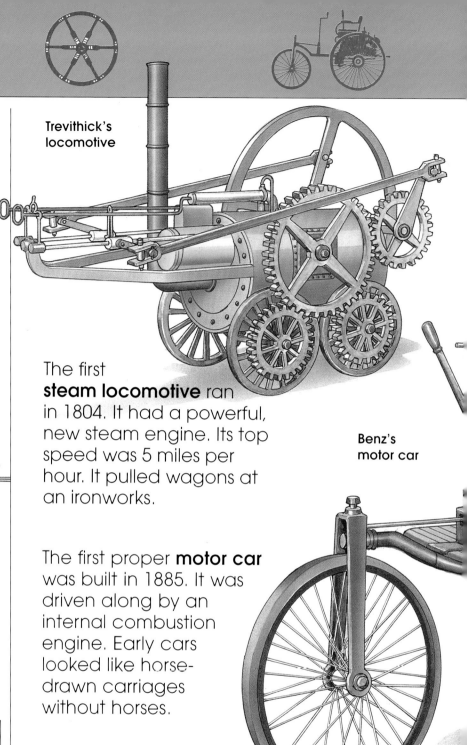

Trevithick's locomotive

The first **steam locomotive** ran in 1804. It had a powerful, new steam engine. Its top speed was 5 miles per hour. It pulled wagons at an ironworks.

Benz's motor car

The first proper **motor car** was built in 1885. It was driven along by an internal combustion engine. Early cars looked like horse-drawn carriages without horses.

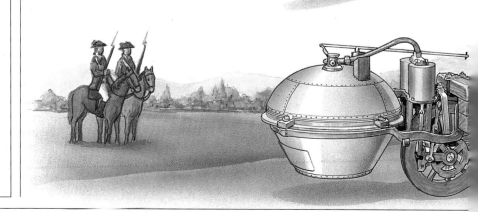

The first **electric locomotive** was demonstrated in 1879 in Berlin.

The pedal **bicycle** was invented in 1839 by a Scottish blacksmith, called Kirkpatrick Macmillan. He only built one machine, which he rode himself.

To make the bicycle go, the rider pushed the pedals backward and forward.

The first **motorcycle** was simply a bicycle fitted with a steam engine. It was built in 1868. The engine was under the saddle.

Steam engines powered tractors, trucks and buses. The first **steam vehicle** was designed to pull military cannons.

Cugnot built his steam tractor in 1769 or 1770. It was slow and quickly ran out of steam.

7

Sea and sailing

A triangular sail, called a **lateen sail,** was invented around 300 B.C. Boats with lateen sails could sail where their crews wanted them to.

A type of boat called a dhow has a lateen sail.

Ships and boats are very old inventions. Archaeologists think that people first made journeys in small boats 50,000 years ago. The boats were very simple canoes carved from tree trunks.

In the 1400s **full rigging** was developed. Full-rigged ships had two or three masts with square and triangular sails.

Ships and boats are not only used for transporting people, but for trade, too. Today, most of the goods traded between different countries are sent by ship.

In the 1400s and 1500s European explorers, such as Christopher Columbus, sailed small full-rigged ships across the oceans.

Archaeologists don't really know when **sailboats** were invented. However, the Ancient Egyptians sailed boats made of reeds along the Nile River over 5,000 years ago.

These reed boats had square sails.

Soon after small portable steam engines were invented engineers built **steam-powered boats**. The *Charlotte Dundas* was built in 1801.

The engine turned large paddle wheels which pushed the boat through the water.

A hydrofoil has wings which lift it out of the water. This means it can go much faster than ordinary boats.

Forlanini hydrofoil

The idea for the **hydrofoil** was thought of in 1881. However, the first hydrofoil was not tested until 1905.

For thousands of years sailors steered using large oars attached to the side of their ship. The **rudder** was invented in China around 700.

Most modern ships are pushed along by a **propeller**. It was patented in 1836 and soon replaced paddle wheels.

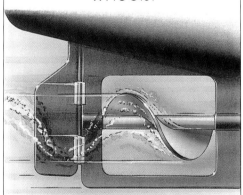

A **Hovercraft** is half boat, half airplane. It skims across sea or land on a cushion of air. The first practical Hovercraft was launched in 1959.

Flying machines

People dreamed of flying like birds for thousands of years before flying machines were made. Many people tried to copy the way birds flew. They tied wings to their arms, but with little success.

Today, there are many different types of aircraft. Every day, millions of people travel around the world in airliners and private aircraft. War planes include small fighters, bombers and huge transport planes.

The first aircraft with wings were **gliders**. Otto Lilienthal made many short glider flights in the 1890s.

The first machine to carry a person into the air was a **hot-air balloon**. It was built by the French Montgolfier brothers in 1783.

The balloon flew 5 miles on its first flight.

The first successful **airship** was built in 1852. In the early twentieth century, airships were popular for transport.

Airships were pushed along by propellers and steered by a rudder.

The Wright brothers built the first **powered aircraft**. They designed and built their own engine because the other engines available were too heavy. Their aircraft first flew in 1903.

The Comet was the world's first **jet airliner**. It made its maiden (first) flight in 1949.

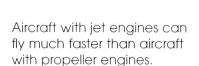

Aircraft with jet engines can fly much faster than aircraft with propeller engines.

The Wright's aircraft was called the *Flyer*. It was pushed along by propellers.

The first successful **helicopter** was built in 1936. It had two spinning rotors. A few years later, a helicopter with a single rotor was built. Most modern helicopters have one rotor.

The Harrier is the most successful VTOL aircraft. The first prototype flew in 1960.

Some aircraft can take off and land straight up and down without needing a runway. They are called **vertical take off and landing aircraft**, or VTOL for short.

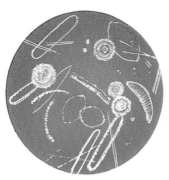

Five hundred years ago people could see only what was visible with their own eyes. Nobody knew how their bodies worked or what was out in space.

When the microscope and the telescope were invented, scientists and astronomers began to discover microscopic cells and millions of new stars.

In 1931, the first **electron microscope** was built. The picture of the object being studied appears on a screen.

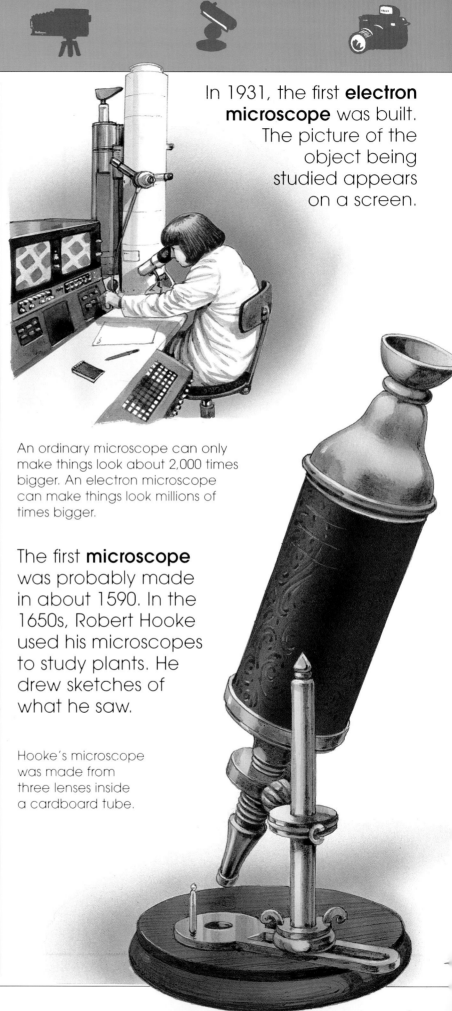

An ordinary microscope can only make things look about 2,000 times bigger. An electron microscope can make things look millions of times bigger.

The first **microscope** was probably made in about 1590. In the 1650s, Robert Hooke used his microscopes to study plants. He drew sketches of what he saw.

Hooke's microscope was made from three lenses inside a cardboard tube.

The first **telescope** was probably made in 1608. The next year Galileo Galilei built his own telescope and used it to study the stars.

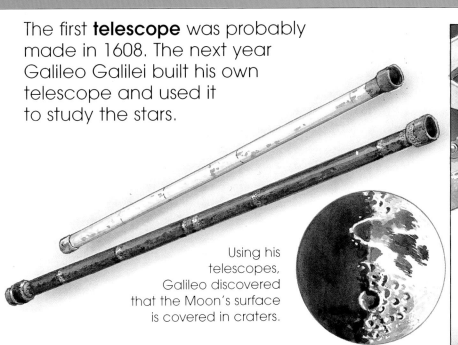

Using his telescopes, Galileo discovered that the Moon's surface is covered in craters.

A **reflecting telescope** uses mirrors instead of lenses. It was first made in 1668. Most telescopes used by astronomers are reflecting telescopes.

Isaac Newton's reflecting telescope

Radio telescopes collect radio waves coming from outer space.

Radio waves from space were first detected in 1931.

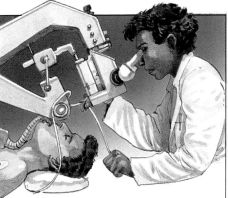

An **endoscope** is a long tube for seeing inside the human body. The first flexible one was built in 1956.

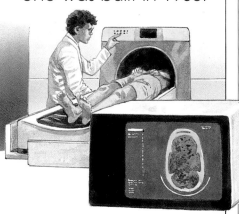

The first **medical scanner** was built in 1971 for looking inside the brain. Scanners for looking at the whole body soon followed.

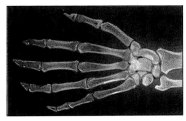

X-rays were discovered in 1895 by Wilhelm Röntgen. They were soon being used to take pictures of human bones.

Weaving a fabric

The first clothes worn by human beings were made from animal skins and fur. Later, people learned to make cloth from other natural materials, such as plants, and still later, from artificial fibers.

Most types of cloth are made on a loom. The loom weaves threads together. Modern looms work automatically. Some weaving is still done on traditional hand looms.

Some materials, such as cotton and wool, have to be spun before they are woven. Spinning makes short fibers into long thread.

People started to spin wool and cotton fibers into thread many thousands of years ago. The first **spinning machine** was like a long spinning top.

The **spinning wheel** was probably invented in India. People started using it in Europe around the year 1300.

A spinning wheel spins and collects thread at the same time.

Around 1767 James Hargreaves invented a machine that he called the **spinning jenny**. It spun thread automatically and made spinning much quicker.

The **loom** appeared around 5000 B.C. The first looms were very simple.

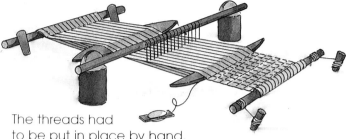

The threads had to be put in place by hand.

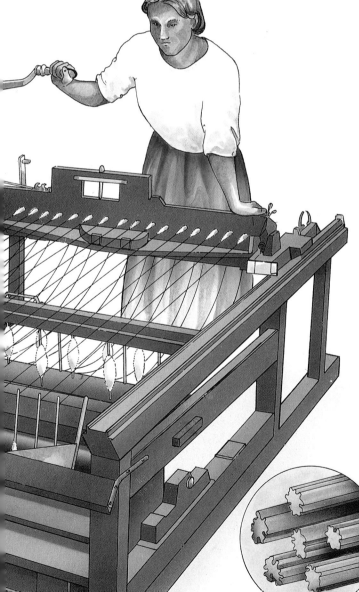

Weaving by hand was very slow. The **flying shuttle** was invented in 1733. It carried the thread from side to side automatically. Before this, it was passed through by hand.

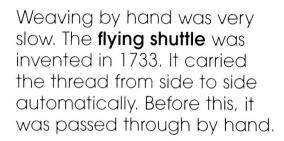

The **Jacquard loom** was invented in 1801. It could weave complicated patterns into the cloth.

The loom was controlled by rows of holes in a long strip of card. Early computers used the same idea.

Rayon fibers, as seen under a powerful microscope.

The first **artificial fiber** was patented in 1892. It was an artificial silk, called rayon.

Thousands of years ago, people did not need to tell the time. They got up when the Sun rose and went to bed when it set. Gradually, as life became more complicated, clocks began to play a larger part in people's lives.

The first clocks were used for waking priests and monks in time for their nightly prayers. Today, clocks seem to rule our lives.

The first clocks were **shadow clocks**. The shadow moved as the Sun moved across the sky. They were invented around 3,500 years ago.

Mechanical clocks were probably developed in Europe during the 1200s. They did not have a face or hands, but rang bells.

The speed of the clock was controlled by a mechanism called an escapement, but it was not very accurate.

In a **water clock**, water drips out of a container so that the level of water inside gradually falls. The Ancient Egyptians were using water clocks about 1500 B.C.

The **pendulum clock** was invented in 1657. It was much more accurate than the clocks before it.

Each swing of the pendulum takes the same amount of time. This keeps the clock running at the same speed all the time.

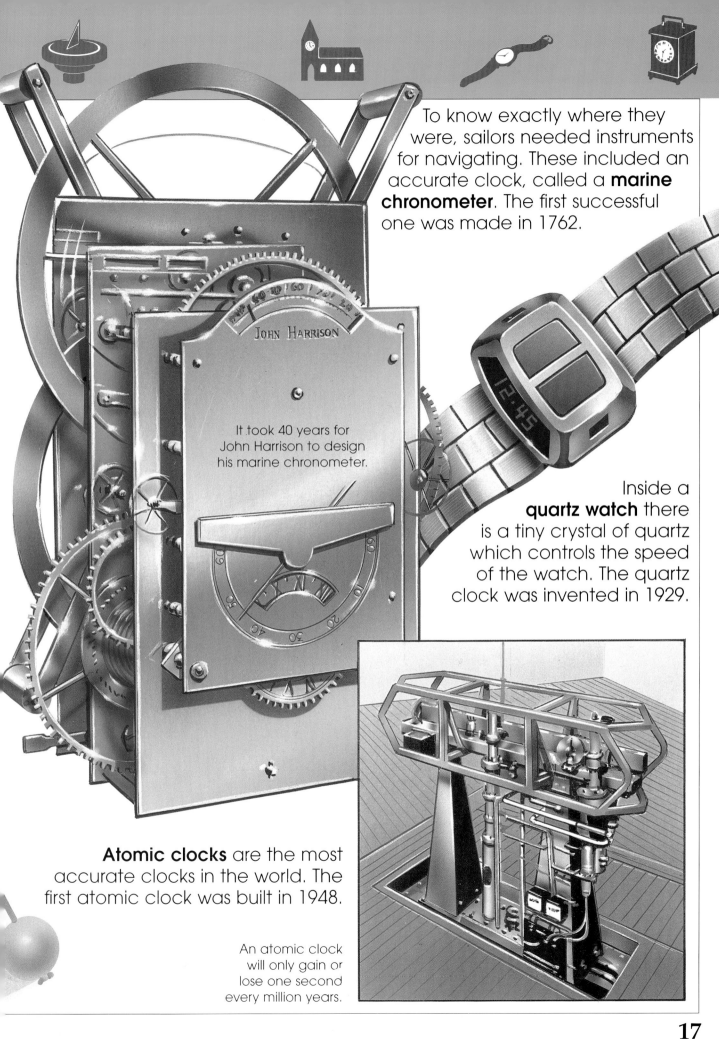

To know exactly where they were, sailors needed instruments for navigating. These included an accurate clock, called a **marine chronometer**. The first successful one was made in 1762.

It took 40 years for John Harrison to design his marine chronometer.

JOHN HARRISON

Inside a **quartz watch** there is a tiny crystal of quartz which controls the speed of the watch. The quartz clock was invented in 1929.

Atomic clocks are the most accurate clocks in the world. The first atomic clock was built in 1948.

An atomic clock will only gain or lose one second every million years.

The Ancient Egyptians used **picture writing**. Each small picture stood for a word or sound. These pictures, or symbols, are called hieroglyphics.

For thousands of years, people did not write anything down. Instead, they passed on information and stories by word of mouth. Shapes and pictures were the first sort of writing.

The first simple **pens** were brushes, or hollow reeds, dipped in ink. The Ancient Greeks used a metal, or bone, stylus to write on soft wax tablets. Later, people used quill pens made from goose feathers.

The end of a goose feather was sharpened and then cut to make a nib shape. To write with a quill, you have to keep dipping the nib in ink.

The first **ballpoint pens** were made in 1938 by Lazlo Biro. When a cheap ballpoint pen runs out, you throw it away. For other pens, you can buy an ink refill with a new ball.

Inside the tip of a ballpoint pen is a tiny steel ball. It rolls around as you write, spreading ink on to the paper.

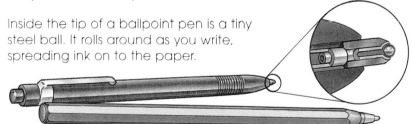

The books that we know today were not made until printing was invented. Until then, every book was copied by hand by people called scribes. Long books took months to copy.

The first **printed book** that still exists was made in China in 868. It is a long roll of paper, and is called the *Diamond Sutra*.

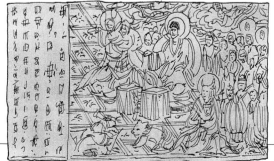

The *Diamond Sutra* was printed by pressing carved, wooden blocks covered with ink on to the paper.

Around 1450, Johannes Gutenberg built the first **printing press**. It could print about sixteen pages of a book every hour.

In 1939, **phototypesetting** was invented. It has now replaced metal type. The words are now typed onto a computer and printed out on photographic paper.

Gutenburg made up words by putting metal letters, called type, together.

Newspapers were first printed in Europe at the beginning of the 1600s. Before then, newspapers were only printed when there was a lot of news.

19

The first machine to record sound and play it back was the **phonograph**. It was invented in 1877 by the American inventor Thomas Edison.

Listening to recorded music is something most people do every day. However, when sound recording was first invented it was a novelty, and nobody took it seriously.

Speaking into the phonograph made a needle move up and down. As the drum went around, the needle made a groove in the tin foil.

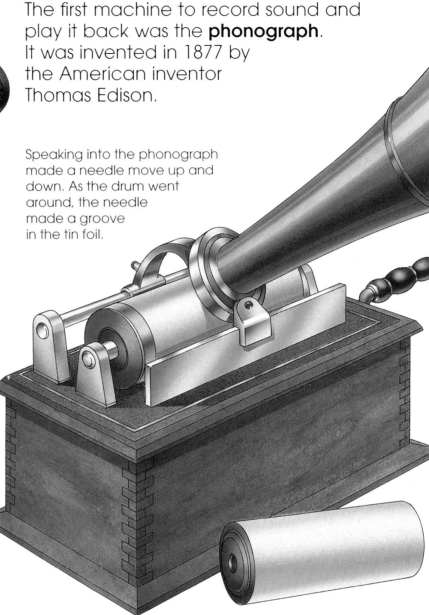

Every so often, a new way of recording sound is invented. Recordings of speech and sounds are also important historical records.

A **tape recorder** records sound as a magnetic pattern on a long strand of tape. The first tape recorder used iron wire. Plastic tape coated with magnetic material appeared in 1935.

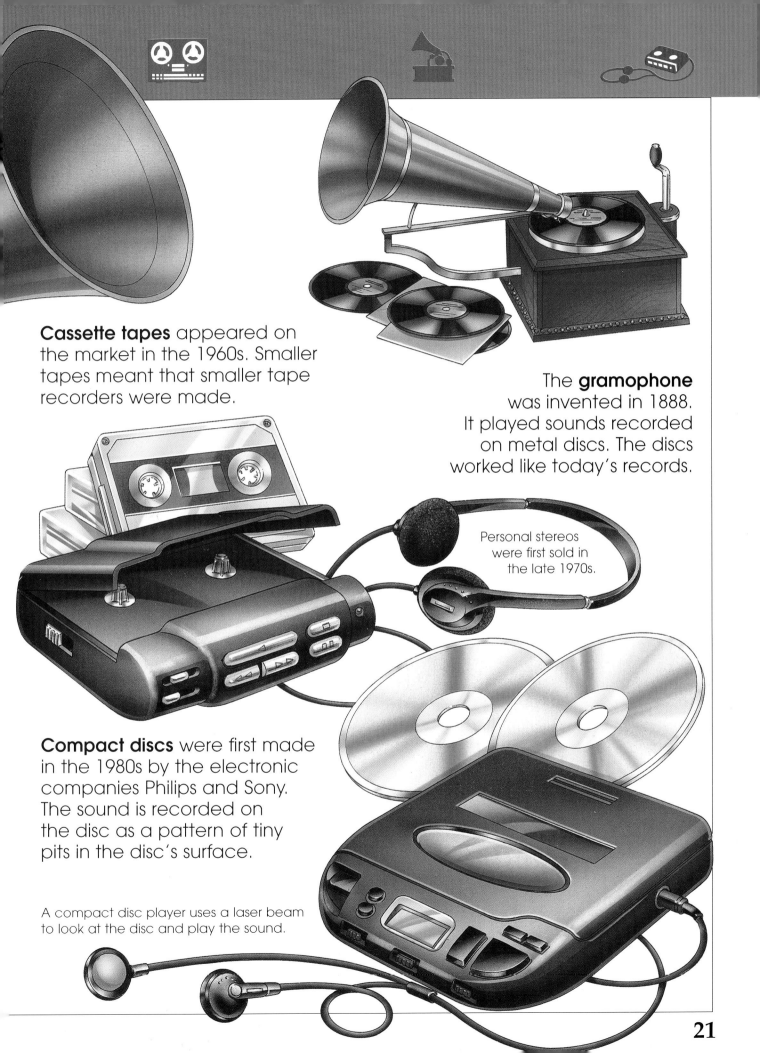

Cassette tapes appeared on the market in the 1960s. Smaller tapes meant that smaller tape recorders were made.

The **gramophone** was invented in 1888. It played sounds recorded on metal discs. The discs worked like today's records.

Personal stereos were first sold in the late 1970s.

Compact discs were first made in the 1980s by the electronic companies Philips and Sony. The sound is recorded on the disc as a pattern of tiny pits in the disc's surface.

A compact disc player uses a laser beam to look at the disc and play the sound.

Until about 200 years ago the only way to send a message was by messenger or by mail. Sometimes, hilltop bonfires were used to send emergency signals.

Today, you can talk on the telephone to friends and relations in almost any part of the world. It takes just a few seconds to dial. Your call might even travel via a satellite in space on the way.

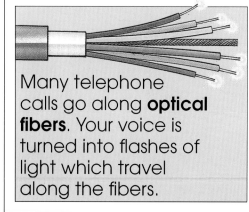

Many telephone calls go along **optical fibers**. Your voice is turned into flashes of light which travel along the fibers.

The **semaphore** system was the first way of communicating over long distances. Semaphore stations were positioned on hilltops, and the message was passed from one station to the next. The system was first used in France in 1794.

The message was shown by moving the arms on top of the semaphore station to different positions.

Messages were tapped with the key as dots and dashes.

The first simple electric **telegraph** was built in 1804. In 1837, Samuel Morse invented **Morse Code**. It was used to send messages by telegraph.

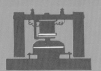

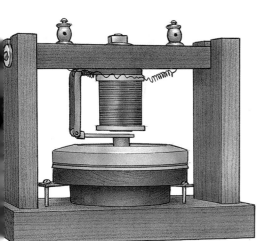

In 1876, Alexander Graham Bell patented the **telephone**. It converted sound into electrical signals. The signals were sent down a wire to another phone and turned back into sound.

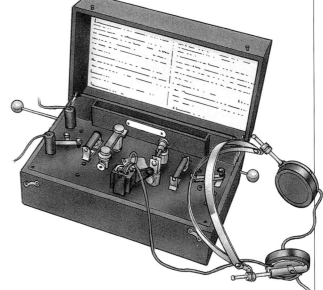

Radio was first used in the 1890s. Sailors used it to send signals to the shore by Morse Code. The first radio program was broadcast in 1906.

An automatic telephone exchange was in operation in 1897. **Electronic telephone exchanges** were built In the 1960s.

The first **telephone exchange** was built in 1878. Only a few people could use it, and it needed a person to operate it.

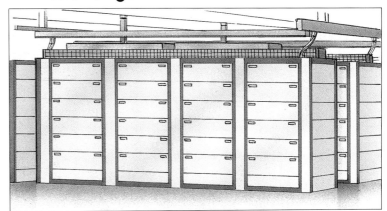

Facsimile machines (fax machines for short) send words and pictures along telephone lines. The idea for fax was first thought of in 1843, but it took until the 1980s for faxes to become common.

Until the 1820s there were no photographs or films. To make pictures of anything, people had to draw or paint them. Taking photographs is a much easier process.

When moving pictures first appeared nobody took them seriously. The machines that made the pictures move were thought of as toys.

Television is part of our everyday lives. We can watch soap operas, films, the news and sports. Thanks to satellites, we can even watch events happening live around the world.

The first **camera** of the type we use today was made by the Eastman company in 1888. It had film that you could send away for processing.

The **kinetoscope** was invented in 1891 by Thomas Edison. You had to look through the top and wind a handle. The film inside lasted only about 15 seconds.

Inside the kinetoscope was a long strip of film with hundreds of pictures on it. Each picture was slightly different from the one before to make an action sequence.

The first **cinema** opened in Paris in 1895. The film was projected on to a screen. The projector worked like the kinetoscope.

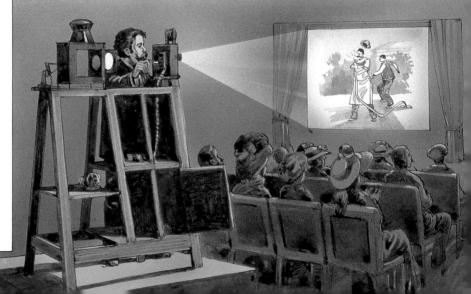

The first time **television** pictures were transmitted by electricity was in 1926. The pictures weren't very good - they were in black and white, wobbly and blurred.

The pattern of light and dark on the picture was made by a spinning disc with holes in it.

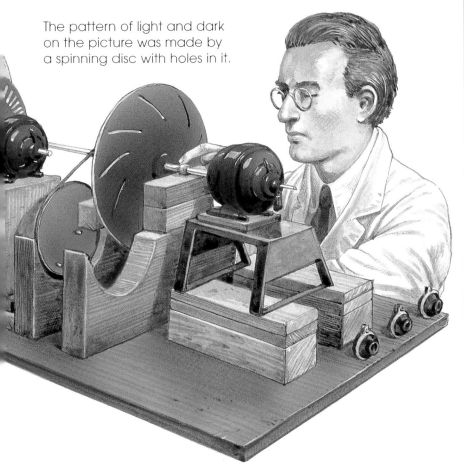

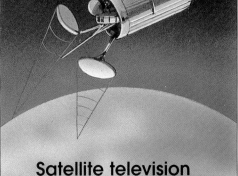

Satellite television receivers that could be installed in homes became popular in the 1980s. The pictures are beamed down from satellites orbiting in space.

In 1928, the first television program was broadcast In America. It was used to test a new **television transmitter**. The pictures were of Felix the Cat™

Color television pictures were first broadcast in 1953.

Video tape and **video recorders** were invented in 1956. Pictures are recorded on videotape just as sounds are recorded on audio tape.

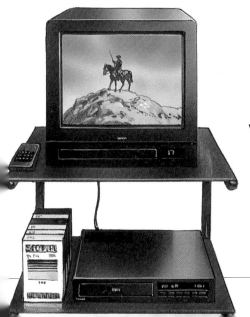

25

Electronic circuits are often used to control and work machines. Computers, televisions and telephones all use electronics. So do some simpler machines, such as washing machines and alarm clocks.

Electronic circuits are made up of electronic components. There are many different sorts of components. One of the most important is the transistor. Its invention meant that electronic circuits could be made much smaller than before.

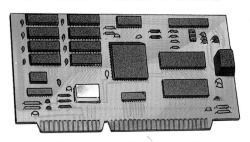

In the 1830s, years before electronics were possible, British scientist Charles Babbage designed a mechanical computer. He called it an **analytical engine**. It was never finished.

The first electronic device was called the **thermionic valve**. It was first made in 1904.

Thermionic valves were used in early radios and televisions.

The first general-purpose electronic **computer** was called ENIAC, which stands for Electronic Numerical Integrator and Calculator. It was built in 1946.

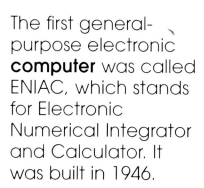

ENIAC used over 18,000 valves and filled a whole room.

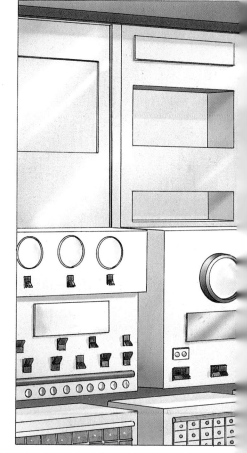

The **transistor** was invented in 1948 by a team of scientists in America. Transistors took over from valves, but were much smaller and cheaper.

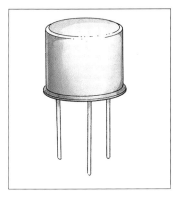

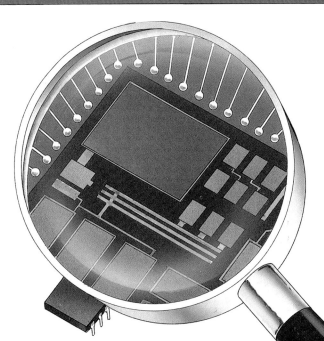

A **silicon chip**, or microchip, as small as a fingernail can contain many thousands of transistors and other electronic components. The first silicon chip was made in 1959.

A silicon chip in a plastic casing

Engineers began to fit more and more components on to a silicon chip. Eventually engineers at Intel built a complete computer on a single chip. This is called a **microprocessor**.

Every personal computer has a microprocessor "brain."

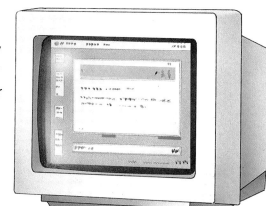

Until the eighteenth century people did all their household chores by hand. There were no washing machines or vacuum cleaners. No one had running water or a flushing toilet either.

The first domestic appliances were mechanical. It was still hard work to operate them. Things really changed when electric motors became cheap to make. Imagine what life today would be like without electricity!

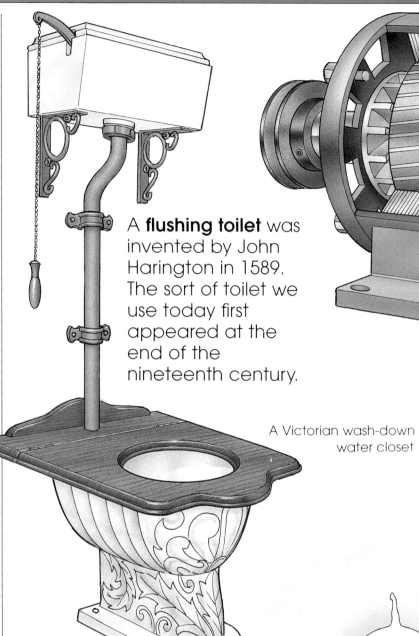

A **flushing toilet** was invented by John Harington in 1589. The sort of toilet we use today first appeared at the end of the nineteenth century.

A Victorian wash-down water closet

Englishman Joseph Swan made a long-lasting **light bulb** in 1878. The next year, Thomas Edison made a similar bulb.

In Edison's light bulb, the electricity flowed through a piece of carbonized bamboo, making it glow.

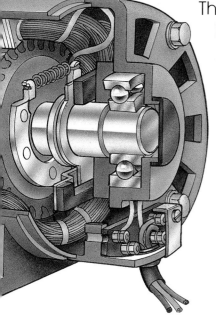

The first **electric motor** was made in 1835. Its power came from a battery because there was no mains electricity at the time.

Operating Booth's Patents

BRITISH VACUUM CLEANER No 11512

Before refrigerators, food was kept fresh in a cool place or boxes lined with ice. The ice had to be replaced as it melted.

The **vacuum cleaner** was patented by Englishman Hubert Booth in 1901. Booth's first machine had to be hired, together with people to operate it.

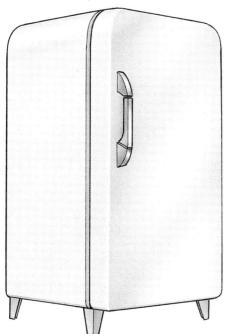

Microwave ovens appeared in the 1950s. They were used by catering companies.

Refrigerating machines were developed at the end of the nineteenth century. It was not until the 1950s, however, that domestic refrigerators became popular.

Microwave ovens cook most foods many times faster than electric or gas ovens.

Weird and wonderful inventions

For every invention that has been a success, there are many more that have been failures. The great age of crazy inventions was the nineteenth century when inventing things became many people's favorite hobby.

Some inventions have no chance of success because their inventor has not understood the scientific principles behind them. Others are simply flights of fancy, designed for fun.

In the fifteenth century, the artist **Leonardo da Vinci** made drawings of many machines, including tanks and flying machines, long before they were actually invented.

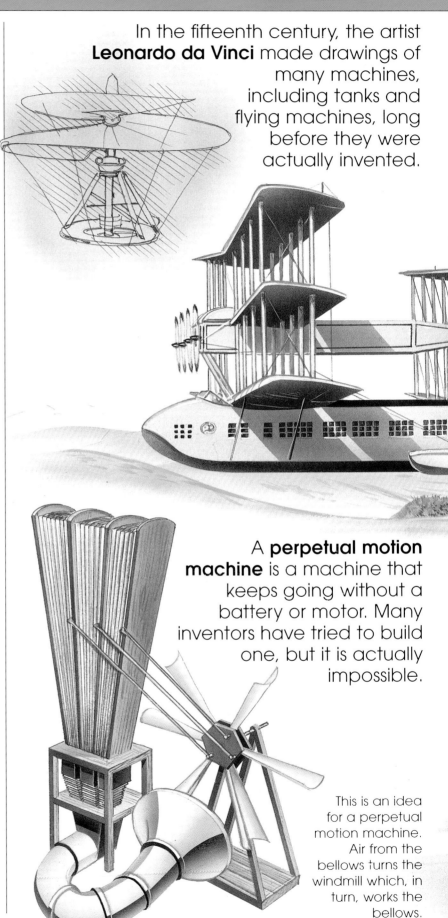

A **perpetual motion machine** is a machine that keeps going without a battery or motor. Many inventors have tried to build one, but it is actually impossible.

This is an idea for a perpetual motion machine. Air from the bellows turns the windmill which, in turn, works the bellows.

The Italian Count Caproni built several huge airplanes. The largest was the **Ca 60**, which had nine wings and eight engines.

The Ca 60 crashed just after take-off on its first flight in 1921.

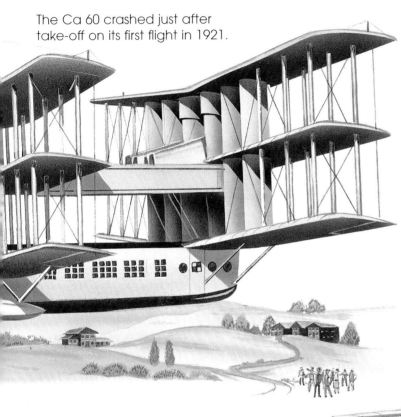

Rube Goldberg was a cartoonist who drew very complicated machines that were completely useless.

The airstrip was never built because it was too expensive.

The *Sinclair C5* was a tiny electric car. It was designed in the 1980s for cheap travel around town. However, most people thought it was too dangerous to drive.

Sir Clive Sinclair

In 1942, English inventor Geoffrey Pyke designed an **iceberg airstrip** on which aircraft could refuel in the Atlantic Ocean.

Airship 10
Analytical engine 26
Artificial fibre 15
Atomic clocks 17

Babbage, Charles 26
Ballpoint pens 18
Bicycle 7
Booth, Hubert 29

Ca 60 31
Camera 24
Cassette tapes 21
Cinema 24
Colour television 25
Compact disc 21
Computer 26

da Vinci, Leonardo 30

Edison, Thomas 20, 24, 28
Electric locomotive 7
Electric motor 29
Electron microscope 12
Electronic telephone
  exchanges 23
Endoscope 13

Facsimile machines 23
Flushing toilet 28
Flying shuttle 15
Full rigging 8

Galileo 13
Gliders 10
Goldberg, Rube 31
Gramophone 21
Gutenburg, Johannes 19

Helicopter 11
Hot-air balloon 10
Hovercraft 9
Hydrofoil 9

Iceberg airstrip 31
Internal combustion
  engine 5

Jacquard loom 15
Jet airliner 11

Jet engine 5

Kinetoscope 24

Lateen sail 8
Light bulb 28
Loom 15

Macmillan, Kirkpatrick 7
Marine chronometer 17
Mechanical clock 16
Medical scanner 13
Microprocessor 27
Microscope 12
Microwave oven 29
Morse code 22, 23
Motor car 6
Motorcycle 7

Newspapers 19
Newton, Isaac 13

Optical fibres 22

Pen 18
Pendulum clock 16
Perpetual motion machine 30
Phonograph 20
Phototypesetting 19
Picture writing 18
Powered aircraft 11
Printed book 18
Printing press 19
Propeller 9

Quartz watch 17

Radio 23
Radio telescopes 13
Reflecting telescope 13
Refrigerating machines 29
Röntgen, Wilhelm 13
Rudder 9

Sailing boats 8
Satellite television 25
Semaphore 22
Shadow clocks 16
Silicon chip 27
Sinclair C5 31
Spinning jenny 14
Spinning machine 14
Spinning wheel 14
Steam engine 5
Steam locomotive 6
Steam machine 4
Steam-powered boats 9
Steam vehicle 7
Swan, Joseph 28

Tape recorder 20
Telegraph 22
Telephone 23
Telephone exchange 23
Telescope 13
Television 25
Television transmitter 25
Thermionic valve 26
Transistor 27
Turbine 5

Vacuum cleaner 29
Vertical take off and
  landing aircraft 11
Videotape 25
Video recorders 25

Water clock 16
Water wheels 4
Wheels 6
Windmills 5

X-rays 13

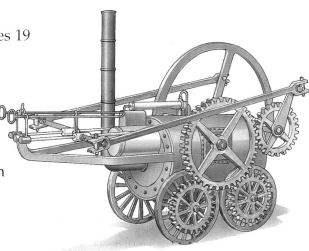